ONE SMALL SQUARE®

Woods

by Donald M. Silver

illustrated by

Patricia J. Wynne

LEARNING
TRIANGLE
PRESS

Connecting kids, parents, and teachers
through learning

An imprint of McGraw-Hill

New York San Francisco Washington, D.C. Auckland Bogotá
Caracas Lisbon London Madrid Mexico City Milan
Montreal New Delhi San Juan Singapore
Sydney Tokyo Toronto

Every plant and animal pictured in this book can be found with its name on pages 38–43. If you come to a word you don't know or can't pronounce, look for it on pages 44–47. The small diagram of a square on some pages shows the distance above the ground for that section of the book.

For Marc Gave

—our editor and dear friend.

We wish to thank Leslie Kimmelman for her skillful styling of the text; Dr. Mary Ellen Holden, Richard Kutner, and Darrel Livesay for their insights into the ways of the woods; Artie Hermann and the staff at Gutenberg Printing, for their exceptional work on this series; and Maceo Mitchell and Thomas L. Cathey for their essential contributions to One Small Square.

Text copyright © 1995 Donald M. Silver.
Illustrations copyright © 1995 Patricia J. Wynne.
All rights reserved.
One Small Square® is the registered trademark of Donald M. Silver and Patricia J. Wynne.

Library of Congress Cataloging Number 97-074147
ISBN 0-07-057933-4

15 16 QTN/QTN 14 13

Whether you are in the woods or at home, always obey safety rules! Neither the publisher nor the author shall be liable for any damage that may be caused or any injury sustained as a result of doing any of the activities in this book.

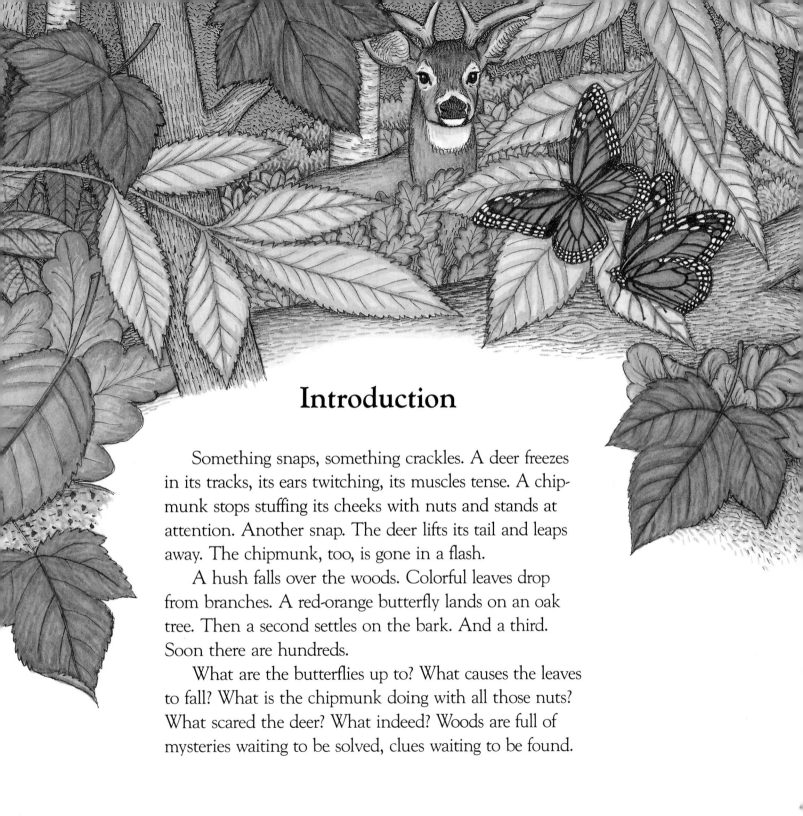

Introduction

Something snaps, something crackles. A deer freezes in its tracks, its ears twitching, its muscles tense. A chipmunk stops stuffing its cheeks with nuts and stands at attention. Another snap. The deer lifts its tail and leaps away. The chipmunk, too, is gone in a flash.

A hush falls over the woods. Colorful leaves drop from branches. A red-orange butterfly lands on an oak tree. Then a second settles on the bark. And a third. Soon there are hundreds.

What are the butterflies up to? What causes the leaves to fall? What is the chipmunk doing with all those nuts? What scared the deer? What indeed? Woods are full of mysteries waiting to be solved, clues waiting to be found.

But it takes a little detective work. And that's where you come in.

You can be a detective! All you need is the simple equipment shown here—and one small square of woods. This book will show you how. There will be hints about what to look for and activities that will help your detective work. If you don't live close to a woods, you can do the activities in your own backyard or in a park near your house where trees and shrubs grow.

No matter where you choose to search for clues, first ask an adult if you may work there. If the place you select is on someone else's land, you must get the owner's permission before you begin. Be sure to tell the adult that you will not harm plants, animals, or the soil. Promise not to explore the woods on your own until the adult feels there is no chance of your getting lost.

With a notebook, an inexpensive magnifying glass, and the other simple equipment shown here, you will soon be on your way to solving some of the woods' many mysteries.

1. Ask an ADULT to find out if hunting is allowed in your woods and, if so, when. The adult should decide if you must WAIT until hunting season ends before you can work in your small square. Hunting season is very dangerous. Hunters wear BRIGHT-ORANGE CLOTHING so they can be seen clearly from far off.
2. Never visit the woods on your own unless an ADULT gives you permission.
3. The EDGE of the woods may be the safest place for you to choose your square. It will be easy for you to reach without getting lost. If you want a square deeper in the woods, you must explore with an ADULT until the adult gives you permission to come and go by yourself.
4. NEVER eat or drink anything you find in the woods.
5. Listen to a WEATHER REPORT before visiting the woods. You don't want to be surprised by a thunderstorm. If you are in the woods and the SKY DARKENS, the air suddenly gets colder, or your hear THUNDER, leave the woods at once and go indoors. NEVER stand under a tree during a thunderstorm. Lightning may strike it, putting you in serious danger.

One Small Square of Woods

Forget the stories you have heard and the movies you have seen. No dragons, evil witches, or trolls have ever lived in the woods. But that doesn't mean that woods are free of danger. Woods can be dark and deep. They can change with the seasons. Before you step inside a woods, STOP! Read the Safety First column on this page. Then you will be ready to find your own small square.

Using a twig, draw a square the size you want in the earth. Make sure your square has at least one tree in it. The square shown here is about four feet (1.2 meters) long on each side—around the size of a four-person elevator.

This square is in a woods where most of the trees lose their leaves in autumn. Yours may be, too. If so, take along some string or ribbon and tie it around the trees in your square. This way, you will be able to find your square even when layers of leaves cover your lines.

The woods await you. Before you know it, you will find clues to how the woods work. These clues will lead to more mysteries and more clues. You will never tire of exploring there—whether you are eight, eighteen, or eighty.

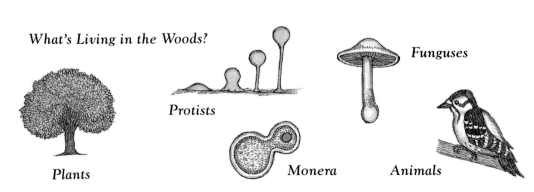

What's Living in the Woods?

Plants

Protists

Monera

Funguses

Animals

Feel how thin a leaf is. It has just a few cell layers. Hold the leaf up to the light. The dark lines you see are veins—tubes that carry food and water. What you can't see are the stomates, openings for the gases moving in and out.

Many hungry woods animals eat plants. In this way, food makers become food.

Waxy covering

Cell

Vein

Stomate

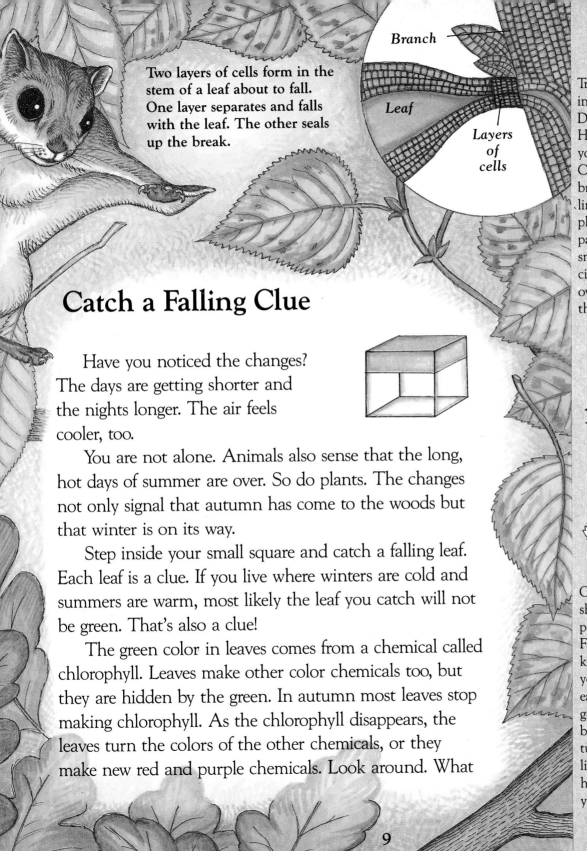

Two layers of cells form in the stem of a leaf about to fall. One layer separates and falls with the leaf. The other seals up the break.

Branch

Leaf

Layers of cells

All Fall Down

Trees that shed all of their leaves in autumn are called deciduous. Deciduous means "falling off." Hundreds of leaves may fall in your small square in autumn. Collect a few, or dozens, and bring them home. Trace the outline of each leaf on paper. Or place a leaf between two sheets of paper. Then, with the paper on a smooth, hard surface, rub a pencil or a crayon back and forth over the upper sheet. Watch as the print of the leaf appears.

Compare the different sizes and shapes. Which leaf tips are pointed? Which are rounded? Fallen leaves are clues to the kinds of trees growing in or near your square. You can identify each leaf by looking in a field guide to trees. A field guide is a book with the names and pictures of plants and animals that live in different places. You may have field guides at home. If not, you can find them at the library.

Catch a Falling Clue

Have you noticed the changes? The days are getting shorter and the nights longer. The air feels cooler, too.

You are not alone. Animals also sense that the long, hot days of summer are over. So do plants. The changes not only signal that autumn has come to the woods but that winter is on its way.

Step inside your small square and catch a falling leaf. Each leaf is a clue. If you live where winters are cold and summers are warm, most likely the leaf you catch will not be green. That's also a clue!

The green color in leaves comes from a chemical called chlorophyll. Leaves make other color chemicals too, but they are hidden by the green. In autumn most leaves stop making chlorophyll. As the chlorophyll disappears, the leaves turn the colors of the other chemicals, or they make new red and purple chemicals. Look around. What

Trees aren't alone in getting ready for winter. Some birds hide seeds to eat later. Others eat to put on fat. Fat is fuel for flying.

Birds use their beaks the way people use knives and forks. The shape and size of a beak provides clues to the kinds of food a bird eats. Match each beak to a bird on this page. Which tears at captured animals? Catches flying insects? Cracks open seeds? Makes holes to probe for hidden insects?

colors do you see? Each kind of tree has its own colors.

Now look up at the leafy roof of the woods. Some leaves are still green. Others have turned color but haven't yet fallen. The green leaves are still making food, using energy from the sun. But something else is going on in the leaves of other colors. Tiny tubes carrying water from the tree's roots to each leaf have become clogged. So have the tubes delivering food from the leaf to the rest of the tree. Slowly each leaf begins to separate from its twig. Feel the end of the leaf stem in your hand. That's where the clogged tubes are. That's where the leaf broke off from its tree.

The leaf you caught is dead. But the tree it came from is very much alive. It is preparing for the coming winter. Before losing its leaves the tree must store food and

10

At least once a year adult birds shed their feathers. Look for fallen feathers in your square. Are they long, stiff flight feathers? Or rounded body feathers? Or soft, downy feathers?

Body feather

Down feather

minerals in its branches, trunk, and roots in order to keep healthy during the cold months. As each leaf falls, a layer of cells seals the break point. This layer stops water from leaking out of the tiny tubes left behind in the tree.

Why do trees shed their leaves in autumn? The answer is right in your hand. On the underside of the leaf are thousands of stomates. The stomates are like miniature mouths. When the "mouths" open, gases from the air can enter the leaf. At the same time, however, water escapes as a gas. Like you, plants need water all the time. In winter, when the ground is frozen, they can't get enough. By losing their leaves, trees lessen the amount of water they need. Go ahead—catch another clue!

Flight feather

Your Woods Notebook

You don't want to forget any of the clues you've discovered in your small square. So jot down in a notebook everything you come across. Draw pictures of plants and animals too. Be sure to record the time of year, the time of day, and the weather for each visit. Refer often to your notebook as you go about solving mysteries.

Everything but Cats and Dogs

Ask at home if you can have an old sheet or tablecloth. Spread it under a tree in your square. Place a rock at each corner to keep the cloth in place. Then check each day to see how many leaves have fallen. What else "rained down" from above? Nuts, berries, bark, feathers, fur, insects? Each is a clue to something going on inside your square. If you move the cloth under a different tree, do the same kinds of things fall?

Hey, Bud!

In your notebook, draw the buds of as many kinds of trees and shrubs as you can reach. Like leaves, buds can help you identify plants. Now look for ring marks —find the ones nearest to the end buds on a branch. The marks were left by scales that fell off of last year's end buds. Measure the distance between the rings and the end buds on the branches. That's how much each branch grew last year. Can you find the ring marks from two years ago? Three? Which year had the most growth?

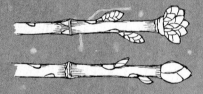

Leaf Life

Hold a plastic tray or dish under a low-hanging branch. Tap the branch once or twice with a stick, letting insects and other leaf life fall into the tray. Draw what you see, using a magnifying glass if necessary. When you are finished, empty the critters onto the ground to give them a chance to stay alive.

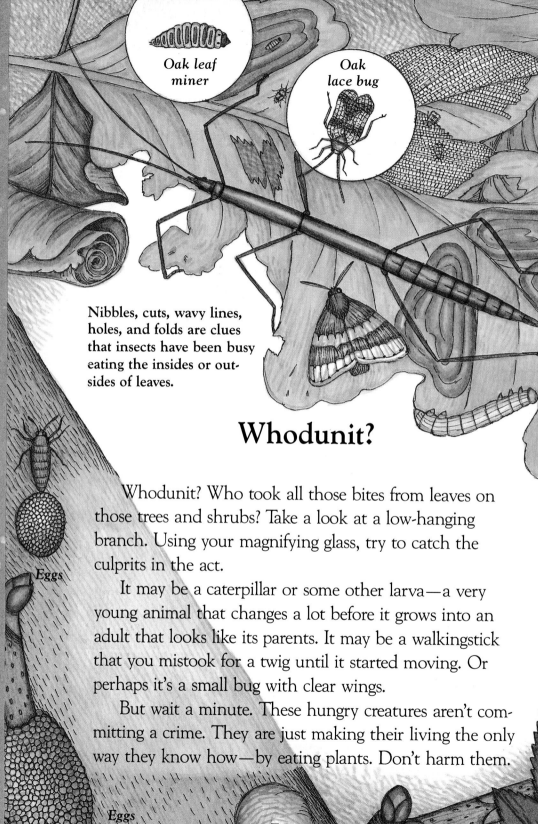

Oak leaf miner

Oak lace bug

Nibbles, cuts, wavy lines, holes, and folds are clues that insects have been busy eating the insides or outsides of leaves.

Eggs

Eggs

Eggs

Whodunit?

Whodunit? Who took all those bites from leaves on those trees and shrubs? Take a look at a low-hanging branch. Using your magnifying glass, try to catch the culprits in the act.

It may be a caterpillar or some other larva—a very young animal that changes a lot before it grows into an adult that looks like its parents. It may be a walkingstick that you mistook for a twig until it started moving. Or perhaps it's a small bug with clear wings.

But wait a minute. These hungry creatures aren't committing a crime. They are just making their living the only way they know how—by eating plants. Don't harm them.

If one of the branches hanging within your reach is from an oak tree, you may notice something strange on its leaves—brown balls. Each ball is called a gall and has at least one larva living inside. When the larva hatched from its egg, it gave off chemicals that caused part of the leaf to swell. The gall is not only the larva's home, but it is also the larva's food. By the time the larva eats its way out of its gall it will be an adult. Adult what? Wasp, fly, mite, worm—their larvas all live and grow in galls.

If you visit your square one morning and discover dark and shriveled leaves, there's only one answer to "Whodunit?" First frost.

Take a close look at the tips of branches with your magnifying glass. You will find buds. Inside the buds tiny leaves and flowers are rolled-up or folded. In the spring they will burst out and grow. Until then, tough scales will protect them from drying out. Look for buds on other parts of branches, too.

Gall

A hole in an acorn is a clue that a weevil may have bored inside and laid eggs. When the eggs hatch, the acorn will become food for the hungry weevil larvas.

13

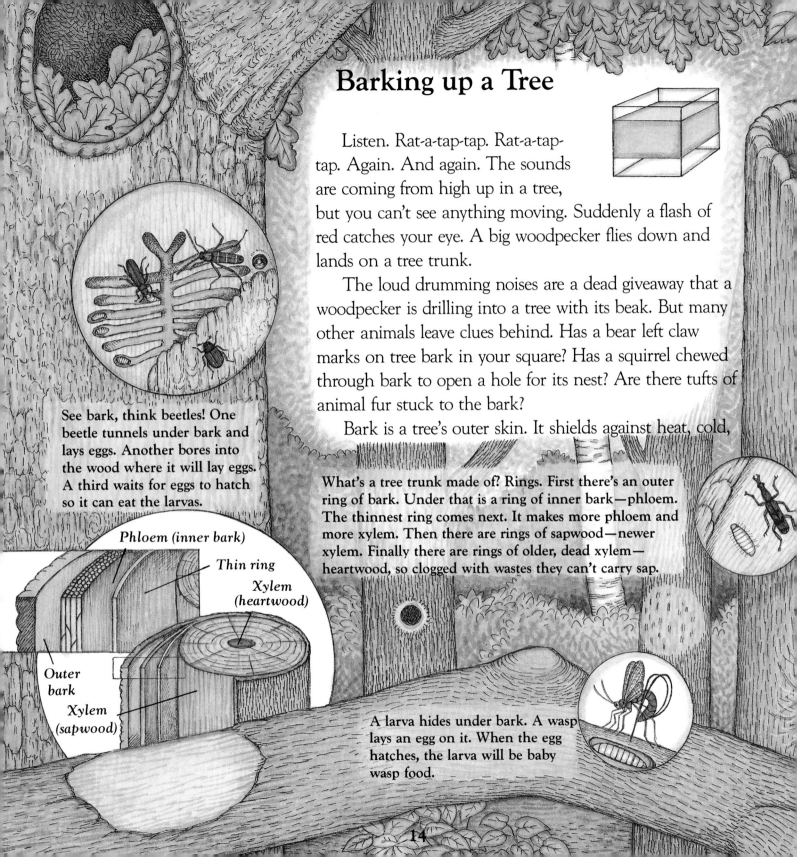

Barking up a Tree

Listen. Rat-a-tap-tap. Rat-a-tap-tap. Again. And again. The sounds are coming from high up in a tree, but you can't see anything moving. Suddenly a flash of red catches your eye. A big woodpecker flies down and lands on a tree trunk.

The loud drumming noises are a dead giveaway that a woodpecker is drilling into a tree with its beak. But many other animals leave clues behind. Has a bear left claw marks on tree bark in your square? Has a squirrel chewed through bark to open a hole for its nest? Are there tufts of animal fur stuck to the bark?

Bark is a tree's outer skin. It shields against heat, cold,

See bark, think beetles! One beetle tunnels under bark and lays eggs. Another bores into the wood where it will lay eggs. A third waits for eggs to hatch so it can eat the larvas.

What's a tree trunk made of? Rings. First there's an outer ring of bark. Under that is a ring of inner bark—phloem. The thinnest ring comes next. It makes more phloem and more xylem. Then there are rings of sapwood—newer xylem. Finally there are rings of older, dead xylem—heartwood, so clogged with wastes they can't carry sap.

Phloem (inner bark)

Thin ring

Xylem (heartwood)

Outer bark

Xylem (sapwood)

A larva hides under bark. A wasp lays an egg on it. When the egg hatches, the larva will be baby wasp food.

14

The beak opens—out comes the woodpecker's long, sticky tongue. There's no escape for a trapped insect.

and drying winds. It is waterproof. It protects a tree from invasion by harmful bacteria and funguses, and from attack by many insects. Flying animals cling to bark when they land on trees. Climbers and crawlers use it as a highway between one part of a tree and another. Bark is a hunting ground for spiders and a hiding place for many small creatures.

Watch a woodpecker hammer into bark. Most likely it is searching for hidden insects. If you see a bird drill rows of holes in the bark, it may be a sapsucker trying to taste the sweet tree sap.

Just under the outer bark are tubes that carry sap. First come tubes called phloem, which forms the inner bark. The sap inside phloem is mostly water mixed with sugars and other foods made in leaves. Phloem carries food *down* from the leaves to the rest of the tree.

There's the Rub

Tape or tie a piece of strong, thin paper to a tree trunk. Slowly rub the flat side of a thick crayon over the paper. Keep all your strokes in the same direction. Make a rubbing of any tree in your square. Is the bark rough? Is it smooth? Does it split? Does it peel into thin strips? Each kind of tree has its own special bark. The patterns of a tree's bark will help you identify it. If you find bark on the ground, you may keep it. But NEVER pull bark off a tree.

Stumped

How old are the trees in your square? The answer lies hidden inside the trunks. Every year a new ring of wood is added to each tree. The ring has two parts, a lighter part that forms in spring, and a darker part that forms in summer. If you find a tree stump, you can figure out how old the tree was when it died —one year for each ring. Are all rings alike? Wide rings mean good growing seasons. Narrow ones mean poor growth that year.

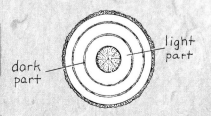

dark part

light part

15

Comings and Goings

What's a scarlet tanager doing in your small square? Or a kingbird? No doubt about it—the fall migration is on. In your notebook make a list of all the migrating guests that drop by your square. A field guide will help with identification. Note the day, time, and weather. How long do the guests stay? What do they eat? Make sure you're in your square for the spring migration, too. In spring, animals migrate back to their summer homes. Do the same kinds of guests stop in your square? Do they look any different? What about next autumn? Check your notebook: Are the dates for the two fall migrations the same?

Even More Clues

Animals leave clues in many places, not just on tree bark. Look for cracked-open nuts on the ground, bitten-off buds on branches, and tracks (like these) in mud or snow.

deer bobcat rabbit

skunk opossum squirrel

This flock of birds flies in a pattern called V-formation. Can you see why?

Then come tubes called xylem, that make up wood. They carry a different kind of sap *up* from the roots to the leaves. This sap is mostly water mixed with minerals. A tree's survival depends on its saps flowing up and down, just as your survival depends on your blood moving continuously through your blood vessels.

As a tree grows, its bark keeps stretching and cracking, then peeling or flaking off. New bark keeps replacing the old from underneath. But no matter how thick, tough, or strong bark is, it can't keep all woods animals out. Beetles tunnel under bark, moths lay eggs in it, and woodpeckers poke holes in it. Deer, porcupines, and worms eat it.

Damage to a tree's bark is like a wound to your skin. To prevent infection and conserve its sap, a tree must seal its wounds. Do your best never to harm any tree's bark.

If you are lucky, you may have a "butterfly tree" in your square. How will you know? You will find it completely covered with red-orange monarch butterflies. The monarchs are resting before they fly to their warm winter homes. In a day or so they will take to the skies, joining millions of birds also escaping months of cold. Perhaps you've already noticed some birds you've never seen before dropping by to snack. Did you suspect something was going on? If so, you were barking up the right tree.

Each monarch butterfly starts life as an egg. Then it hatches into a caterpillar. Inside a hard case called a chrysalis, the caterpillar turns into a butterfly. After mating, adult butterflies lay eggs—and the cycle starts again.

Adult

Egg

Chrysalis

Larva

This grosbeak won't stay for winter. The blue jay will.

Ovenbird

2

Yellow warbler

1

Monarch butterfly

Redstart

4

Blackpoll warbler 5

Match the number next to each bird or butterfly with the same number on the map. Or follow the arrows. Either way, you will find the warm winter home of each animal. The journey between a summer and a winter home is called migration.

17

Talk about disguises! Which is the spanworm, which is the branch? Talk about camouflage! Where does the owl end and the tree trunk begin?

Sudden silence in the woods is a clue that a predator may be on the prowl. Heed the animals' warning and be careful.

As long as leaves stay green, green insects can hide out in the open without being seen.

The tiny eyes of this moth's head are the only ones that are real. But the spots on its wings look just like the eyes of a larger animal. Some predators are tricked and don't attack.

18

Nobody Said

Nobody said detective work would be easy. As more leaves fall, you make more and more noise approaching your small square. If you don't want to scare away the squirrel searching for acorns, the chipmunk filling its cheeks with seeds, or the deer munching on leaves, you'd better move as slowly and as quietly as possible. Even then, you can get only so close before these plant eaters will flee to safety.

Do you blame them? You could be a predator—an animal that wants to eat them. Like everywhere else in nature, woods are full of predators that eat the animals that eat plants.

Imagine you are a predator. How would you catch your prey? Would you hide, then ambush it? Build a web or some other trap? Suddenly swoop down from a branch high up in a tree? Watch the hunters in your small square. Do any of them capture a meal in a way you never thought of?

Nobody said hunting prey was easy. Plant eaters are always alert for danger. They often have tricks to prevent the predators from even noticing them. Some run and hide when they see, hear, or smell anything that could be a hunter. Others freeze in place—their body colors blend-

Poison gland

The spider's bite poisons the insect trapped in its web. The poison paralyzes the prey and turns its body into a liquid that the spider then sucks up.

If you have trouble finding the mantis, you're not alone. So will mantis eaters. And so will the insects the mantis is quietly waiting to ambush.

19

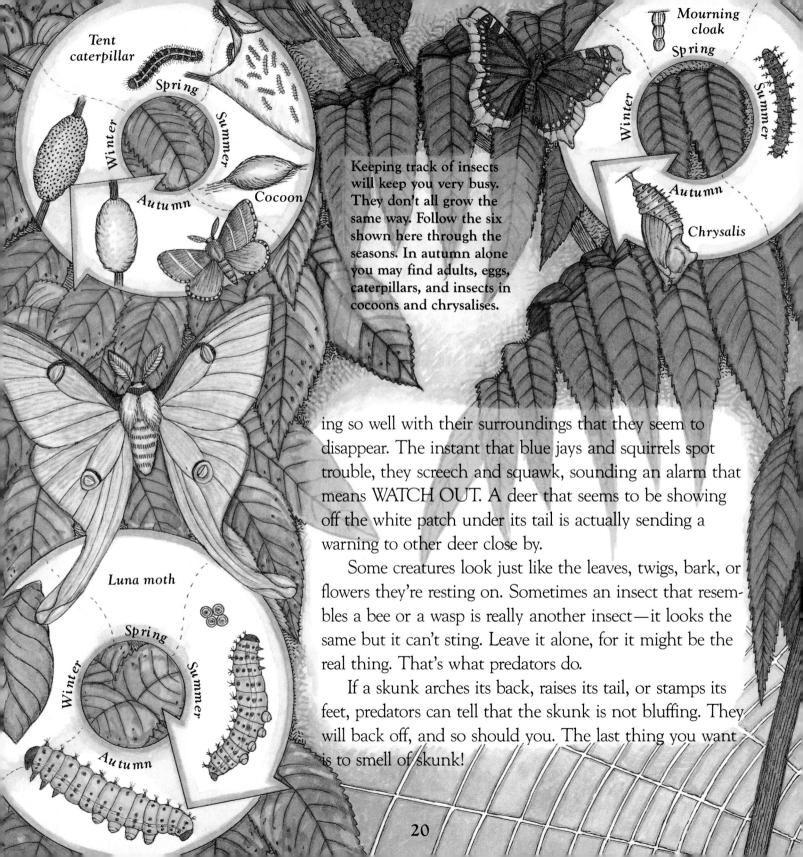

Keeping track of insects will keep you very busy. They don't all grow the same way. Follow the six shown here through the seasons. In autumn alone you may find adults, eggs, caterpillars, and insects in cocoons and chrysalises.

ing so well with their surroundings that they seem to disappear. The instant that blue jays and squirrels spot trouble, they screech and squawk, sounding an alarm that means WATCH OUT. A deer that seems to be showing off the white patch under its tail is actually sending a warning to other deer close by.

Some creatures look just like the leaves, twigs, bark, or flowers they're resting on. Sometimes an insect that resembles a bee or a wasp is really another insect—it looks the same but it can't sting. Leave it alone, for it might be the real thing. That's what predators do.

If a skunk arches its back, raises its tail, or stamps its feet, predators can tell that the skunk is not bluffing. They will back off, and so should you. The last thing you want is to smell of skunk!

Fall cankerworm

Spring
Summer
Autumn
Winter

Try to figure out how insects stay alive at different stages of their lives. Does their diet change as they grow? Do their colors or patterns change? Do they have wings to help them escape or catch dinner?

Milkweed bug

Spring
Summer
Autumn
Winter

Caterpillars don't look at all like the butterflies or moths they eventually become. But young milkweed bugs are already very much like their parents.

Stag beetle

Spring
Summer
Autumn
Winter

For every animal you spot, there are many more nearby. They are hidden so well that you may not suspect they are there. Soon you will begin to figure out the different ways animals trick one another. You will no longer be startled when a moth opens its wings to show off what looks like two huge eyes. When a caterpillar swells up one end of its body to make fake eyes appear, you will understand what it is doing.

For woods animals, hide-and-seek is no game. It is a life-or-death struggle—eat or be eaten. One animal loses its life so that another may live. But don't be surprised to discover that predators also have shapes and colors to help them hide. Or that look like leaves, twigs, flowers, and bark. After all, nobody said that nature takes sides.

Catch another falling clue. This time the clue is the fruit of a maple tree. It has "wings" that spin in the wind like helicopter blades, and it carries a cargo of two seeds.

If you see a bear cub feasting on berries, STAY AWAY, no matter how cute and playful it looks. Its mother may be nearby. She will attack anything that gets too close to her cub.

Get Carried Away

Clues are disappearing from the woods. The birds are in on it. The fox, the mouse, and the bear are too. So are you, but you probably don't know it. That is, unless you notice burrs stuck to your clothes when you return home from your square.

Carefully shake off some of the burrs and look at them under your magnifying glass. See the hooks or spines? They stick easily to clothes—and also cling to the fur and feathers of animals that brush against them.

Burrs are fruits. They form from flowers, and they contain seeds that can grow into new plants. You know that apples and pears are fruits, but you may not have

Square Fruit

It's a good thing most plants make a lot of fruits, some with a lot of seeds. That way there's a better chance at least some seeds may escape being eaten or harmed and grow into new plants. Hunt for fruits in your square. Did the fruit fall from a plant or did some animal carry it in to the square and drop it? Did it ride the wind? If a fruit is split open, try to count how many seeds it has. Draw both the fruit and seeds in your notebook. If you know the name of the plant they came from, all the better. Check each fruit for bite marks. You may be able to tell if a big or a small animal has been nibbling on it. The fruits you find may look delicious but DO NOT eat them—some may be poisonous to people. Next time you eat a fruit at home, collect some of its seeds and look at them under your magnifying glass. Which "vegetables" are actually fruits—because they contain seeds?

realized that tomatoes, berries, and nuts are fruits, too. They all contain seeds.

Like you, animals may carry fruits out of your square. Some of the fruits may drop in soil where there are fewer plants already growing. There the seeds inside the fruits may have a better chance of sprouting than they would in your crowded square.

Plants with berries also depend on animals. Their colorful, tasty fruits attract berry eaters. Migrating birds fill up on the energy-rich berries. Foxes and bears fatten up on them, too. Hours or days later, some of the seeds in the berries may come out unharmed in the animals' droppings. Where? Perhaps close by; perhaps far away.

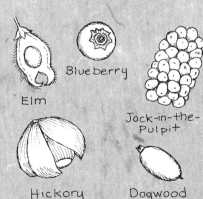

Elm
Blueberry
Jack-in-the-Pulpit
Hickory
Dogwood

Acorn weevil

With a twist and a turn of her snout, an acorn weevil cuts a hole in an acorn. She eats nutmeat and lays eggs, leaving enough food for her larvas to feed on.

The acorn moth lays an egg in a hole in an acorn. When the caterpillar hatches, it covers the hole with its web to keep out other acorn eaters.

Acorn moth

In a Nutshell

There may be a world at your feet—in a nutshell. If there are acorns in your square, pick a few up off the ground. Do you see any cracks or holes? A weevil may have cut a hole in the acorn while it was still on its oak tree. A moth caterpillar may have chewed through the shell to eat out tunnels of nutmeat.

On the ground tiny beetles, springtails, and snails squeeze inside those holes and cracks to nibble away a share of the food. Funguses find their way in to soak up all they can. They, in turn, attract mites and other fungus feeders. A centipede may hunt the creatures busy snack-

Drop an acorn in water. If it floats, its insides have already been eaten.

Filbert worm moth caterpillar

Acorn moth caterpillar

24

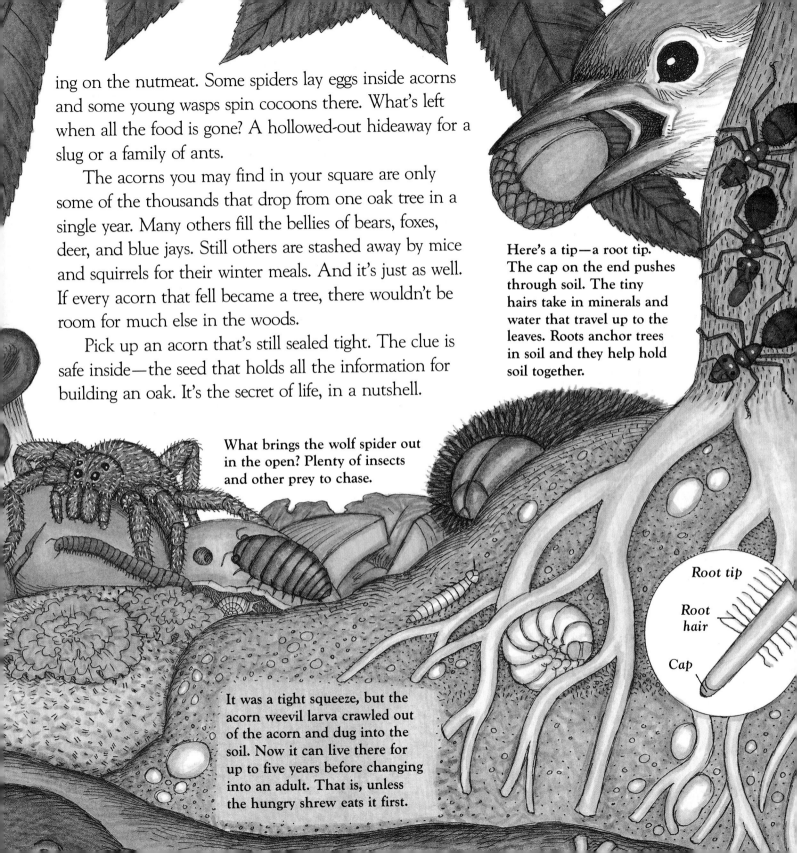

ing on the nutmeat. Some spiders lay eggs inside acorns and some young wasps spin cocoons there. What's left when all the food is gone? A hollowed-out hideaway for a slug or a family of ants.

The acorns you may find in your square are only some of the thousands that drop from one oak tree in a single year. Many others fill the bellies of bears, foxes, deer, and blue jays. Still others are stashed away by mice and squirrels for their winter meals. And it's just as well. If every acorn that fell became a tree, there wouldn't be room for much else in the woods.

Pick up an acorn that's still sealed tight. The clue is safe inside—the seed that holds all the information for building an oak. It's the secret of life, in a nutshell.

Here's a tip—a root tip. The cap on the end pushes through soil. The tiny hairs take in minerals and water that travel up to the leaves. Roots anchor trees in soil and they help hold soil together.

What brings the wolf spider out in the open? Plenty of insects and other prey to chase.

Root tip

Root hair

Cap

It was a tight squeeze, but the acorn weevil larva crawled out of the acorn and dug into the soil. Now it can live there for up to five years before changing into an adult. That is, unless the hungry shrew eats it first.

Wanted: Dead and Alive

There's a dead tree in the small square. It's been lying on the forest floor for years. And yet...it's too alive for any detective to ignore. Dead and alive? It's one mystery that's easy to solve!

As soon as the tree fell, beetles began to tunnel under the bark. Water seeped in. Funguses and bacteria invaded and started to soften and break down the wood inside.

Look at the tree now. It is riddled with tunnels and full of cracks. Ants and termites nest within. Mosses and mushrooms grow from it. The tree is alive with snails and sowbugs, salamanders, spiders and centipedes—making

Mites

Springtail

Protists

Bristletail

Roundworm

Bacteria

What's under the fallen leaves in your square? Last year's leaves? Do they feel damp or dry? Are any whole?

Slime mold

Many creatures live among the fallen leaves. You can see some of them under the magnifying glass. Bacteria and most protists are invisible except under a microscope.

26

their living feeding, hunting, and hiding.

Meanwhile bacteria and funguses are causing the dead tree to slowly rot. But more than the fallen tree will decay and disappear. So will last year's leaves which litter the forest floor. And the animal droppings—and pods, galls, and dead animals. Bite by bite they will be eaten by insects, worms, and other litter feeders. Bit by bit they will be broken down into minerals and other nutrients by bacteria, protists, and funguses. These recyclers return the minerals and nutrients to the soil, keeping it fertile. Without recyclers, trees and other plants could not keep growing. Yes, there's a dead tree in the small square, and it helps the woods stay alive.

What killed this tree? It could have been lightning, strong winds, disease, or just plain old age.

Unlike termites, carpenter ants don't eat wood. Instead, they chew out tunnels from the wood for their nest.

27

Let It Rain, Let It Spore

After a rain, mushrooms may pop up in your square. Like all other funguses, mushrooms lack chlorophyll and so they cannot make food. Most get their food from dead plants and animals. Funguses don't make flowers or seeds either. They reproduce by means of spores. Collect a few umbrella-shaped mushrooms and carefully carry them home. (Remember: NEVER eat mushrooms you have found. They may be poisonous.)

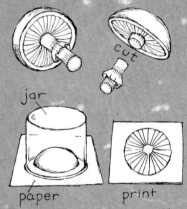

Ask an adult to help you cut each stem close to the caps, as shown. Turn each cap over on a piece of paper. Cover with a jar or bowl to protect the cap and leave overnight. Now remove the bowl and gently lift up the cap. Millions of spores may have fallen out of each one. Draw the pattern of the spores in your notebook. Do all the patterns look alike? Shake the spores off outside. Is a print left on the paper? What do you think scatters spores in the woods?

Why is winter cold? Look to see how high the run rises in the sky. How many hours of daylight are there?

You wear a coat, a scarf, and gloves in winter. Birds and mammals keep warm with fluffed feathers or thick fur.

Most adult insects were killed by the frost. But some are alive under bark.

28

In Winter's Grip

The squirrel did not budge from its nest while the snow fell and the icy wind howled. The squirrel stayed put when the temperature plunged. But when hunger struck, the squirrel wasted no time. Out it came into the cold to sniff the snow. Sound like a "nutty" thing to do? But wait! The squirrel dug right down and pulled out an acorn it had buried weeks ago.

You won't see many animals in your square when the woods are in winter's grip. But at least a few creatures—squirrels and some kinds of birds—can cope with the cold and still find food to eat.

Where are the other animals? Those that stayed in the woods are in places where they will not freeze: in tree holes, under bark, inside a fallen log, or down in the earth, blanketed by leaf litter and the snow. Most will not move or eat until spring; instead they'll live off their body fat. The chipmunk, however, wakes from its sleep now and then to nibble on seeds it stored in its burrow during the warmer weather. It may even surface for a snack before returning to its snug underground hideaway.

Be sure to visit your square during the winter. With all you can discover, it's really not as nutty as it sounds.

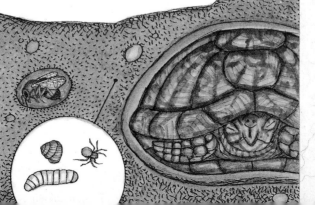

Sure, snow is cold. It slows you down—and animals, too. But it also protects seeds, roots, soil, and animals in their burrows from icy winds.

In Shape

See how well lit the woods are, now that all the leaves are gone. You can see every branch on every tree. Draw different tree shapes in your notebook. Some trees have trunks that divide again and again. Others rise straight up, their smaller branches growing out from the single trunk.

Measuring Up

For this activity you will need a friend and a stick. Ask your friend to stand next to a tree in your backyard. You then stand back about 25 paces from the tree and hold the stick at arm's length. Line up the top of the stick with the top of your friend's head. Move your hand to the place on the stick that lines up with your friend's feet. How many times can you fit the marked-off part of the stick into the height of the tree? Multiply that number by your friend's height—and the answer will be close to the tree's height. Are the trees in your small square taller or shorter than the tree you measured?

Pollen

Stamen

Pistil

Where do most seeds come from? Flowers. Pollen grains from male flower parts, called stamens, must reach female flower parts in the pistil of the same kind of flower, in order for seeds to form.

In a Hurry

Before all the snow melted, before the night air lost its chill, a surprise grew in your small square. It was only one flower. But it was also a clue that spring was about to break winter's grip on the woods.

With the first warm days came more flowers in a hurry to soak up the sun and show off their colors. Insect eggs started hatching, cocoons split open, and animals underground began to stir from their long sleep.

One by one, buds on bushes and shrubs burst open. There were tiny leaves inside some, flowers in others, and both leaves and flowers in still others. Soon buds on trees were adding more food-making leaves to those already back in business.

Most plants need the help of birds, bees, butterflies, and moths to make seeds. When these animals feed on a flower's sweet nectar, pollen sticks to them. At the next flower they visit, some pollen drops off. This process is called pollination.

30

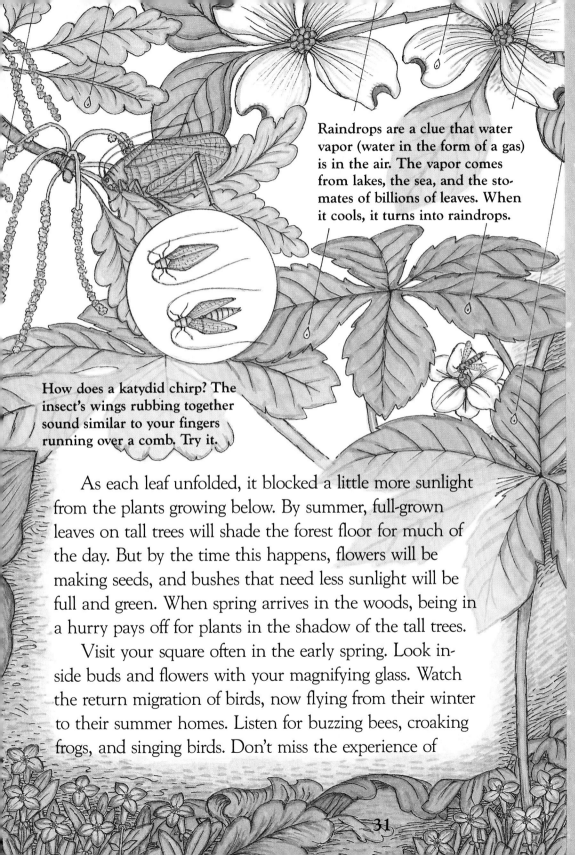

Raindrops are a clue that water vapor (water in the form of a gas) is in the air. The vapor comes from lakes, the sea, and the stomates of billions of leaves. When it cools, it turns into raindrops.

How does a katydid chirp? The insect's wings rubbing together sound similar to your fingers running over a comb. Try it.

As each leaf unfolded, it blocked a little more sunlight from the plants growing below. By summer, full-grown leaves on tall trees will shade the forest floor for much of the day. But by the time this happens, flowers will be making seeds, and bushes that need less sunlight will be full and green. When spring arrives in the woods, being in a hurry pays off for plants in the shadow of the tall trees.

Visit your square often in the early spring. Look inside buds and flowers with your magnifying glass. Watch the return migration of birds, now flying from their winter to their summer homes. Listen for buzzing bees, croaking frogs, and singing birds. Don't miss the experience of

Soil Searching

Check the leaf litter. Have recyclers been busy turning last fall's leaves into a dark, nutrient-rich mush called humus? Dig down into the soil, too. Is there a dark layer of topsoil? A lighter-colored middle layer, the subsoil? A rocky third layer (substratum) where new soil is forming? In which layer do soil animals live? Look for clues. Be sure to replace the soil when you are through.

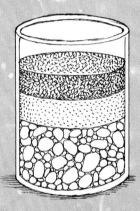

Taking Turns

It's time to look for acorns again —those that have just started to sprout on top of the soil. Notice in which direction the roots and the young stem point. Turn a sprouting acorn on its side. What happens to the roots and the stem? Check each day.

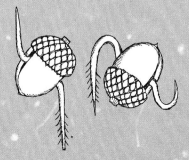

31

You've seen two of these birds in this book. Which ones? Did they look different in the fall?

Look for bird nests low in trees. Not much wind blows there, so the eggs stay safe. Don't you harm the eggs either!

What's inside this egg? A baby bird and the food it needs to grow and hatch.

Downy woodpecker

Be sure to tell an adult if anything bites you. Ask an adult to help you check for ticks on your skin. Some tick bites can make people sick.

Deer tick

Poison ivy

seeing the woods come alive with feeding, growing, building, tunneling, hunting, hiding creatures—all in a rush to find mates.

So what's the hurry? Do a little detective work and you will find out. Keep an eye on the birds. They must build nests, lay eggs, and sit on them to keep them warm until hatching. Then the birds must spend whole days searching for food to feed their babies. The babies need enough time to grow strong, learn to fly, leave the nest, and hunt for their own meals—all before the chill of autumn signals the return of winter.

Watch, too, the moth just out of its cocoon. By the time the warm weather ends, it must mate and then lay eggs that will hatch into caterpillars. These, in turn, eat and eat, and grow and grow, until each one spins its own cocoon.

As the days get longer and warmer, it may seem as if something is going on in every nook and cranny of your square. Don't be in a hurry. Spring is the perfect season in the woods for a detective hot on the trail of clues.

Poison ivy, poison oak, or poison sumac may grow in your square. They can cause itchy, burning rashes. So learn what these plants look like and do your best to avoid them.

Ball

Thread

Spore

Some mushroom spores grow underground threads. When threads from two spores join, a ball forms, pops out of the earth, and grows a stalk with an umbrella-shaped cap full of new spores.

33

Woodsorama

Take a shoebox and measure its length and height. Cut a piece of paper for the background wall, about 1/4 inch (6 mm) shorter than the box height and about 4 inches (10 cm) longer than its length. On the paper draw some trees and other woods plants. Place the picture in the box and tape each side to the front. The picture will curve.

On separate sheets of paper color other plants, animals, and rocks—each with a flap. Cut out each picture, bend its flap, and glue or tape it to the box. Create a woods diorama for every season. Use the dioramas to explain to family and friends about the mysteries you have solved in the woods.

If these young raccoons don't pay attention, they may end up as owl food.

A Tale Told by a Clue

The long, hot summer days no longer bother you. You've discovered that shady woods are a great place to beat the heat. The problem is that the leaves not only block out the sun but also keep you from seeing what many of the woods animals are up to. Don't worry. Animal clues will help you figure out what's going on.

Early one morning you find something fuzzy, brown, and pickle-shaped on the ground under a tree, something that wasn't there the day before. It's not alive, but it looks as if there's a bone or two inside. Using twigs you gently pick it apart—it's full of bones, teeth, and fur. What could it be?

Soft edges on an owl's wing feathers muffle its flight. It strikes silently, without warning.

Some plants grow and even flower in the shade.

To solve this mystery, you must imagine the woods in the dark of night. Animals are on the prowl for their next meal. A shrew munches on a ground beetle. A mouse feasts on seeds. Suddenly, without making a sound, an owl swoops down and flies off with the mouse in its strong, curved claws. The predator lands on a branch and swallows its prey whole. But it cannot digest the mouse's bones, teeth, and fur. So the bird coughs them up into the fuzzy clue you found—a pellet.

This tale is true. You *can* tell from an owl pellet exactly what the owl has been eating. You really can unlock some of the secrets of the woods at night from clues you find in the morning.

Lots of baby animals mean lots of food for predators. As long as some of each kind of animal survives to have young, woods remain full of life.

Owl pellets

A forest fire can burn down hundreds of trees, kill or injure animals, and destroy their homes. It will take many, many years for plants to return. A forest fire can be started by a bolt of lightning —or a careless person. Make sure you are always careful in the woods.

Clued In

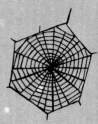

Spider in web

You've guessed, haven't you? Everything you find in your square—from the tops of the tallest trees, to the bushes and shrubs, to the forest floor, and even underground—is a clue to how the woods work.

The woods are not merely the trees, but also all the other plants and all the creatures large or small that live there. All the different ways they have of making a living add up to how the woods work. Animals and plants need each other and the recyclers to stay alive. They also need you to do nothing that will harm them or where they live.

Your detective work has only just begun. Year after year the woods keep changing. There will always be new clues for you to find. And the more you understand about the living woods, the better you will be at uncovering them. So when the leaves under your feet crackle and fallen twigs snap, *you* may be the clue to the mystery of why a deer freezes in its tracks and a chipmunk leaps away.

Gray birch

Marginal wood fern

Jack-o'-lantern mushroom

Marble salamander

Small Square in Autumn

Can you match each living thing with its outline?

Blue jay

Black and white warbler

Chickadee

Katydid

Gray fox

Monarch
butterflies

Flying
squirrel

Hornbeam

American
chestnut

Hairy
woodpecker

Nuthatch

Polyphemus
moth
larva

White
oak

Red
maple

Pileated
woodpecker

Gray
squirrel

Pignut
hickory

Flowering
dogwood

Scarlet
tanager

Rose-breasted
grosbeak

Moss

Winterberry

Sumac

Cottontail
rabbit

Deer
mouse

Five-lined
skink

Titmouse

Blueberry

Ovenbird

Luna
moth

Short-tailed
shrew

Cardinal

Carolina mantis

Fawn mushroom

Artist's fungus

Eastern
spotted
newt

Earthworm

Chipmunk

Gray
tree
frog

All these wood animals are vertebrates—animals with backbones.

Fur is a clue that you are looking at a mammal; feathers, a bird. Reptiles have a dry, scaly skin, but an amphibian's skin is thin, moist and scaleless.

Opossum

Raccoon

Short-tailed shrew

Mammals

Chipmunk

White-tailed deer

Long-tailed weasel

Deer mouse

Striped skunk

Black bear

Cottontail rabbit

Silver-haired bat

Flying squirrel

Gray squirrel

Gray fox

Fox squirrel

Red-breasted grosbeak

Black and white warbler

Screech owl

White-breasted nuthatch

Redstart

Chickadee

Red-shouldered hawk

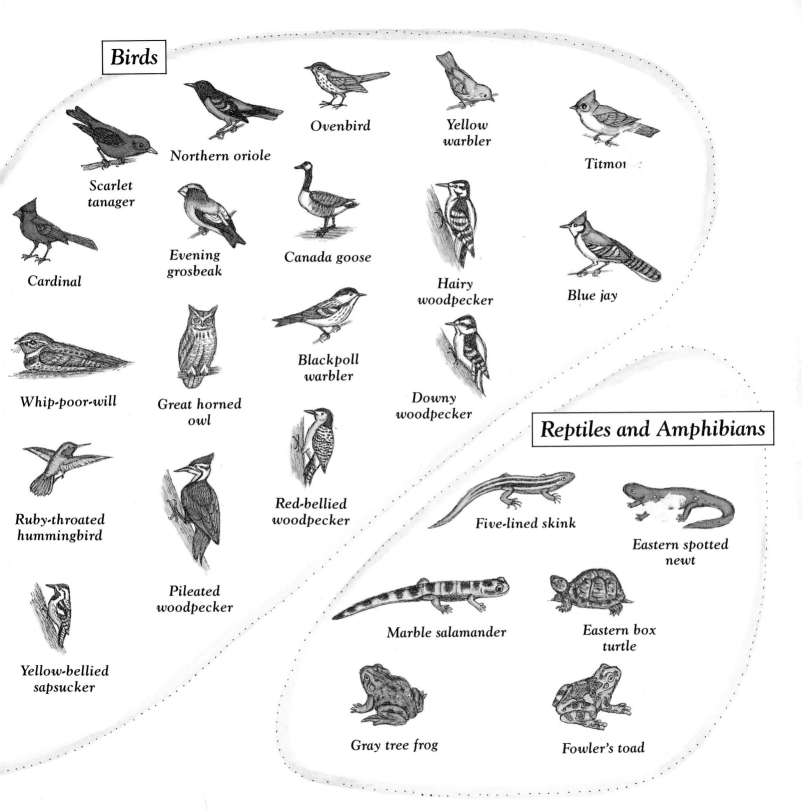

Birds

Scarlet tanager

Northern oriole

Ovenbird

Yellow warbler

Titmouse

Cardinal

Evening grosbeak

Canada goose

Hairy woodpecker

Blue jay

Whip-poor-will

Great horned owl

Blackpoll warbler

Downy woodpecker

Ruby-throated hummingbird

Red-bellied woodpecker

Pileated woodpecker

Yellow-bellied sapsucker

Reptiles and Amphibians

Five-lined skink

Eastern spotted newt

Marble salamander

Eastern box turtle

Gray tree frog

Fowler's toad

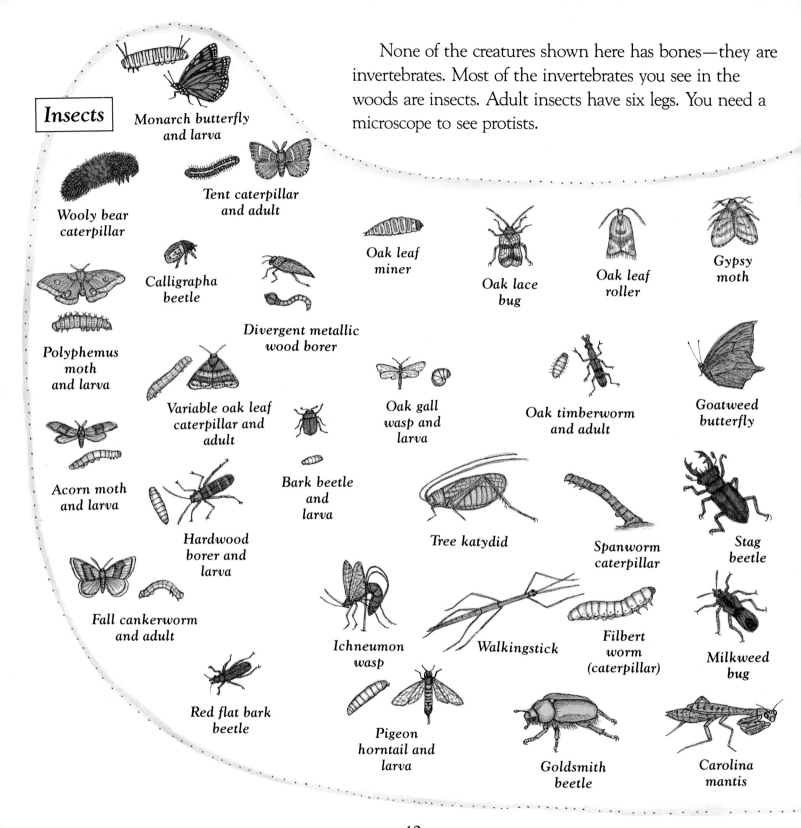

Insects

None of the creatures shown here has bones—they are invertebrates. Most of the invertebrates you see in the woods are insects. Adult insects have six legs. You need a microscope to see protists.

Monarch butterfly and larva

Tent caterpillar and adult

Wooly bear caterpillar

Calligrapha beetle

Divergent metallic wood borer

Oak leaf miner

Oak lace bug

Oak leaf roller

Gypsy moth

Polyphemus moth and larva

Variable oak leaf caterpillar and adult

Oak gall wasp and larva

Oak timberworm and adult

Goatweed butterfly

Acorn moth and larva

Hardwood borer and larva

Bark beetle and larva

Tree katydid

Spanworm caterpillar

Stag beetle

Fall cankerworm and adult

Ichneumon wasp

Walkingstick

Filbert worm (caterpillar)

Milkweed bug

Red flat bark beetle

Pigeon horntail and larva

Goldsmith beetle

Carolina mantis

40

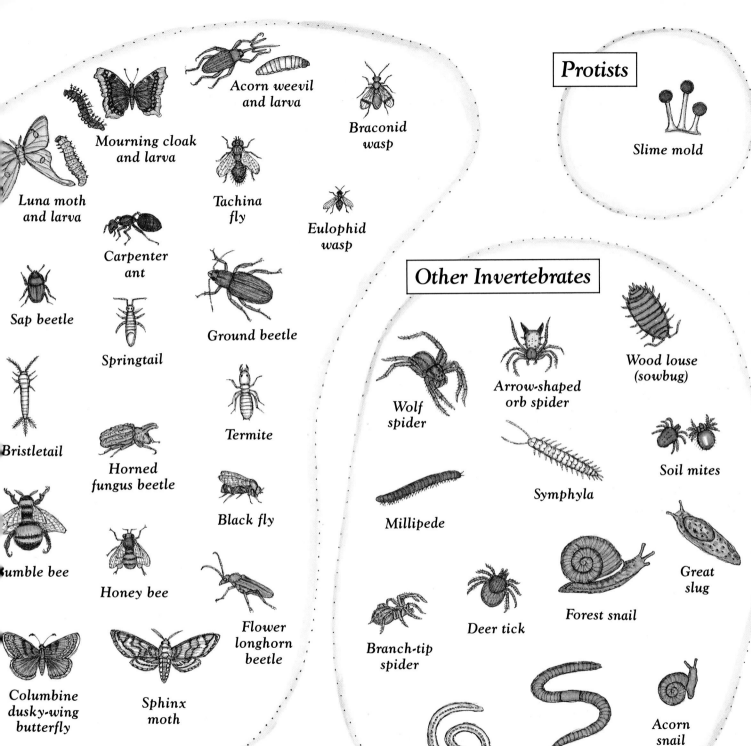

Luna moth
and larva

Mourning cloak
and larva

Acorn weevil
and larva

Braconid
wasp

Tachina
fly

Eulophid
wasp

Sap beetle

Carpenter
ant

Springtail

Ground beetle

Bristletail

Horned
fungus beetle

Termite

Bumble bee

Honey bee

Black fly

Columbine
dusky-wing
butterfly

Sphinx
moth

Flower
longhorn
beetle

Protists

Slime mold

Other Invertebrates

Wolf
spider

Arrow-shaped
orb spider

Wood louse
(sowbug)

Millipede

Symphyla

Soil mites

Branch-tip
spider

Deer tick

Forest snail

Great
slug

Roundworm

Earthworm

Acorn
snail

The green color of plants is a clue that they can capture energy from the sun and use it to make food. You can identify plants by their shape, size, and leaves.

Funguses often look like plants, but they cannot make food. You can see monera only under a microscope.

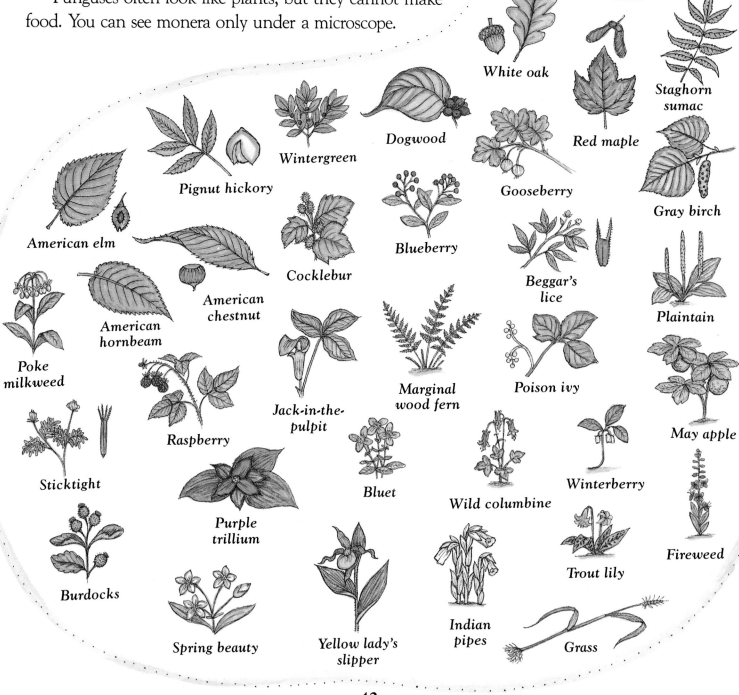

Pincushion moss

White oak

Red maple

Staghorn sumac

Dogwood

Gooseberry

Gray birch

Wintergreen

Pignut hickory

Blueberry

Beggar's lice

American elm

Cocklebur

Plaintain

American chestnut

American hornbeam

Jack-in-the-pulpit

Marginal wood fern

Poison ivy

May apple

Poke milkweed

Raspberry

Sticktight

Bluet

Wild columbine

Winterberry

Fireweed

Purple trillium

Burdocks

Spring beauty

Yellow lady's slipper

Indian pipes

Trout lily

Grass

42

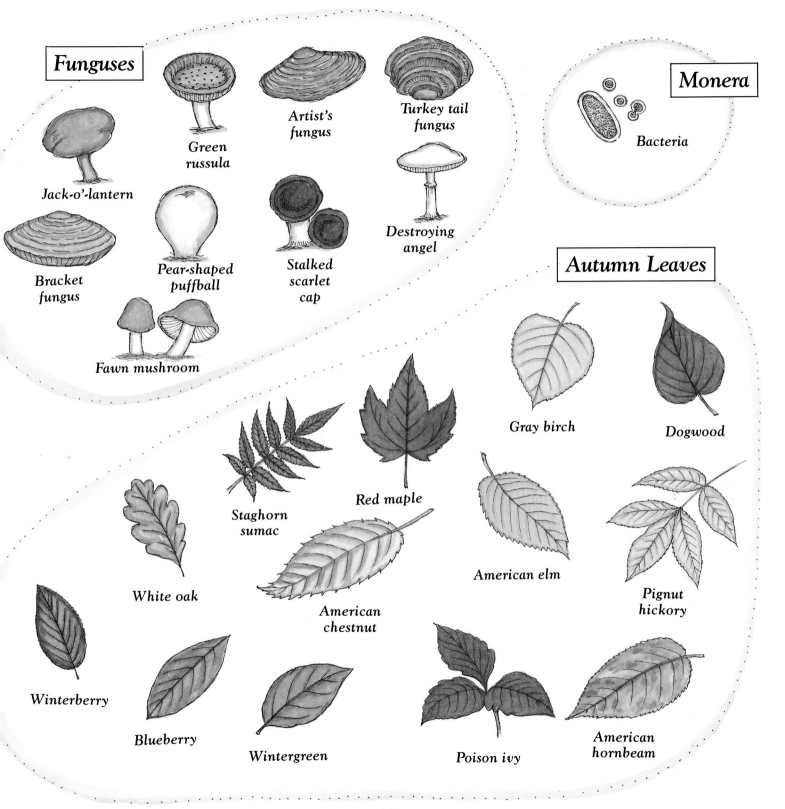

Funguses

Jack-o'-lantern

Green russula

Artist's fungus

Turkey tail fungus

Bracket fungus

Pear-shaped puffball

Stalked scarlet cap

Destroying angel

Fawn mushroom

Monera

Bacteria

Autumn Leaves

Gray birch

Dogwood

Staghorn sumac

Red maple

American elm

Pignut hickory

White oak

American chestnut

Winterberry

Blueberry

Wintergreen

Poison ivy

American hornbeam

Index

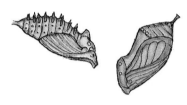
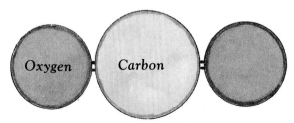

Index

Roof

Shrubs

Floor

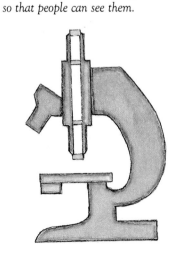

Index

migration 16, 17, 23, 31
mineral 11, 16, 25, 27
mite 13, 24
monera (muh-NEER-uh) 6, 42.
    Creatures made up of one cell
    that doesn't have a nucleus
    (control center).
moss 26

moth 16, 18, 21, 24, 30, 33
mouse 22, 25, 35
mud 16
muscle 3
mushroom 26, 27, 33

N
nectar 30
nest 14, 26, 27, 29, 32, 33
night 9, 30, 35
notebook 5, 11, 12
nut 3, 6, 11, 22
nutmeat 24, 25. *The part inside*
    a nut that animals eat.
nutrient (NOO-tree-int) 27, 31.
    Any part of food that living
    things must have to build cells
    or to use as a source of energy.
nutshell 24, 25

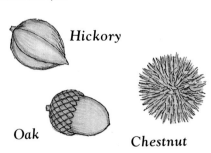

Hickory

Oak

Chestnut

O
oak 3, 12, 13, 24, 25

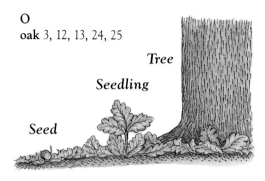

Tree

Seedling

Seed

old age 27
owl 18, 34, 35
owl pellet 35
oxygen 8

P
phloem (FLOH-im) 14, 15
photosynthesis
    (foh-toh-SIN-thuh-siss) 8
pistil 30
plant eater 19
pod 27
poison 19, 22
poison ivy 33
poison oak 33

poison sumac 33

pollen 30
pollination 30
porcupine 16
predator (PRED-uh-tur) 18, 19
    20, 35
prey 19, 25, 35. *Animal hunted*
    or caught for food by a
    predator.
protist 6, 26, 27, 40. *A living*
    thing usually of one cell that
    has a nucleus (control center).

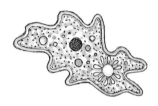

R
raccoon 34
rain 27, 31
recycler 27, 31, 36
reptile 38
root 10, 11, 16, 25, 29, 31
rotting 27

S
salamander 26
sap 14, 15, 16
sapsucker 15
sapwood 14
scale 12, 13, 38

Butterfly

Reptile

Tree

Index

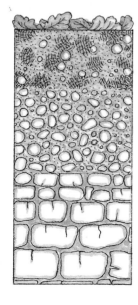

Further Reading

To find out more, look for the following in a library or bookstore:

Golden Guides, Golden Press, New York, NY

Golden Field Guides, Golden Press, New York, NY

The Audubon Society Beginner Guides, Random House, New York, NY

The Audubon Society Field Guides, Alfred A. Knopf, New York, NY

The Peterson Field Guides, Houghton Mifflin Co., Boston, MA

Reader's Digest North American Wildlife, Reader's Digest, Pleasantville, NY

Eyewitness Books, Alfred A. Knopf, New York, NY

Look in an art supply store or in the library for books on how to draw plants and animals. If you like to sketch and paint outdoors, here are some things you'll find handy:

- paper
- number 2 pencil
- paintbrush
- bottle of black ink
- black drawing pen
- tray of watercolors
- eraser
- plastic bottle for water
- stiff cardboard or clipboard to draw on

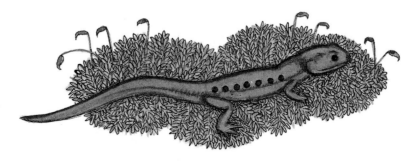